Earth's Detectives

# ALL ABOUT CRYSTALS

BY REBECCA STORM

# CONTENTS

| | | | |
|---|---|---|---|
| WHAT ARE CRYSTALS? | 4 | COLLECTING CRYSTALS | 20 |
| HOW DO CRYSTALS FORM? | 6 | COMMON CRYSTALS | 22 |
| CRYSTAL SHAPES | 8 | ICONIC GEMSTONES | 24 |
| WHICH CRYSTAL? | 10 | UNIQUE CRYSTALS & GEMSTONES | 26 |
| AMAZING CRYSTAL PLACES | 12 | RECORD BREAKERS | 28 |
| WHAT ARE GEMSTONES? | 14 | TRUE OR FALSE? | 30 |
| HOW WE USE CRYSTALS & GEMSTONES | 16 | GLOSSARY & INDEX | 32 |
| GEMSTONE DETECTIVES | 18 | | |

Words in **BOLD** can be found in the glossary.

Copyright © 2025 Hungry Tomato Ltd

First published in 2025 by Hungry Tomato Ltd
F15, Old Bakery Studios, Blewetts Wharf, Malpas Road, Truro, Cornwall, TR1 1QH, UK.

No part of this publication may be reproduced, stored in a retrieval system, or transmitted in any form or by any means, electronic, mechanical, photocopying, recording, or otherwise, without prior written permission of the copyright owner.

A CIP catalogue record for this book is available from the British Library.

ISBN 9781835690826

Printed in China

Discover more at
www.hungrytomato.com

Picture credits:
Abbreviations: m-middle, t-top, l-left, r-right, bg-background.

Shutterstock: 25 (all gemstones), 20 (all birthstones); Agnieszka Gaul 12br; AKaiser 30b; Albert Russ 9ml; Aleksandr Pobedimskiy 11 (corundum); Alexey Boldin 16br; Bjoern Wylezich 9br, 22tl; Byjeng 14mr (ruby), 14mr (emerald); Cagla Acikgoz 26tl; ch123 5br; Cagla Acikgoz 4mr; Chase D'animulls 23tl; Christian Vollmert 12ml; COULANGES 30tr; Dan Olsen 27mr; DmitrySt 11(diamond); Edward Westmacott 28br; EugenePut 15tr; Finesell 31tr; Fokin Oleg 11 (fluorite)Gilmanshin 29mr; GTW 7br; Hayati Kayhan 15b (pearl); Ihor Bondarenko 16tl; Jack Dagley Photography 29tl; JaneMoon 14mr (red corundum); Jirik V 14br; ju_see 4b; Kokophotos 26bl; KrimKate 11 (calcite), 11 (apatite); luchschenF 17BL; Luis War 13b; Merrillie Redden 15br; Michael Tubi 17tl; Myriam B 26mr; NickKnight 29bl; Pesh Siri 31ml; Peter Hermes Furian 8mr, 23bl; photo-world 30b; pixelrain 18m, 19m; sakkmesterke 7bl; Sebastian Janicki 8bl, 11 (quartz), 27bl, 30b; SergeUWPhoto 15ml; Somjit Chomram 25bl; Stone36 5m; TimebirdArt 13t; TR_Studio 11 (topaz); Tyler Boyes 4bl (olvine); Risto Raunio 7t; Roy Palmer 22mr; vvoe 4ml, 9tr, 11 (talc), 14ml (beryl), 15m (coral stone), 15t (amber), 22bl, 23mr, 30b; Wirestock Creators 11 (gypsum), 31br; YesO58 Montree Nanta 6mr, 6br. NASA: http://www.thatcrystalsite.com/. http://photojournal.jpl.nasa.gov/catalog/PIA10615 28m.

Every effort has been made to trace the copyright holders, and we apologise in advance for any unintentional omissions. We would be pleased to insert the appropriate acknowledgements in any subsequent edition of this publication.

# WHAT ARE CRYSTALS?

Crystals are shiny, solid materials. They can be different shapes, sizes, and colours. Crystals have smooth faces and straight edges that fit together in a repeating pattern.

Crystals are found in many places on Earth. Some form in caves deep underground, others form inside stones on the surface of the Earth.

Some crystals are **BRIGHTLY COLOURED**, like this green chrysoprase...

Others are **SHINY AND METALLIC**, like this pyrite...

Others are **SO CLEAR** you can almost see through them, like this clear quartz!

**CRYSTALS ARE ALL AROUND US.** Foods like salt and sugar are made of tiny crystals. We even have crystals in our bodies – our bones are made of millions of very small crystals.

We also use crystals to make things like clocks and games consoles! Sometimes we shape and polish them, and use them in things like jewellery. Shaped and polished crystals are called gemstones.

If you look at salt under a microscope, you can see its crystalline structure more clearly.

Snowflakes are formed when ice crystals join together. They change shape as they fall, meaning every snowflake is **unique**!

# HOW DO CRYSTALS FORM?

The process of crystals forming is called crystallisation. This can happen in different places and different ways, depending on the material that is turning into a crystal.

All rocks are made of at least one **mineral**. Often minerals grow into hard, shiny objects with flat surfaces – crystals! Many rocks and crystals form underground.

Most of Earth's crystals are formed when hot, **molten** rock cools and hardens to create **igneous rocks**. If molten rock cools quickly, the crystals in it are tiny. If it cools slowly, the crystals can be much bigger.

**Granite is an igneous rock.**

Deep under Earth's surface, there is a lot of heat and **pressure**. This can change igneous and **sedimentary rocks** into **metamorphic rocks**. When the rocks change, the crystals in them change too. When granite changes to gneiss, it no longer looks speckled, but stripey.

**Gneiss is a metamorphic rock.**

Crystals can form in different situations on Earth's surface too.

Water that cools down naturally slowly turns into ice crystals. These ice crystals are present in glaciers, icebergs, and even snow.

Some crystals form from volcanic lava. As lava cools, it turns into igneous rock. Crystals that form in igneous rock are angular, unlike crystals in other rocks which are worn down.

Some crystals form when water **evaporates** from a mixture. When seawater evaporates, salt gets left behind and forms into crystals as it dries.

# CRYSTAL SHAPES

Crystals can be many shapes and sizes. They can have lots of surfaces, or just a few. The surfaces can be different shapes and joined in different ways, too.

## SYMMETRY IN CRYSTALS

The surfaces of crystals are called faces. When crystal faces join together, they sometimes make a shape that looks the same from all sides. These crystals are symmetrical and often called "perfect crystals".

### HALITE

This is halite. Its crystals often have six square faces that form like cubes.

### FLUORITE

Some crystals, like this fluorite, can be made of eight triangular faces. They look like two pyramids stuck together!

# UNUSUAL CRYSTAL SHAPES

Some crystals are non-symmetrical and have much more irregular shapes. They still look very impressive though! Here's a few of the different shapes they can be.

## MICA

This is mica, one of the crystals that forms inside the rock, granite. Mica forms in very flat, thin crystals.

## GYPSUM

Gypsum crystals can grow to be incredibly long and thin, which can make them really fragile. They look a bit like needles.

## HAEMATITE

Haematite often forms in round, bubbly shapes! The crystals inside are so small you can't see them with the naked eye.

# WHICH CRYSTAL?

There are many different ways of working out what type of crystal something is. As well as the shape, you can look at the colour and test its hardness.

## COLOURFUL CRYSTALS

Crystals come in many different colours. Some crystals have more than one colour! This can help you work out what type of crystal it is.

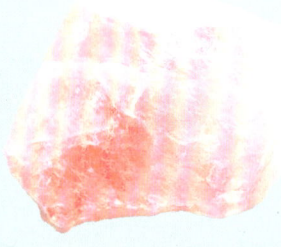

Rose quartz gets its name from its colour. It is always a very pale pink.

Sapphires are a deep blue. They are often formed into gemstones and used for jewellery.

Topaz comes in many colours. When it is gold with a hint of red like this, it is called "imperial topaz".

Malachite is a very intense green. This is why it used to be used to make paint.

Find out more about different crystal colours on pages 22-27.

# CRYSTALS HARD AND SOFT

We use the **Mohs Hardness Scale** to test the hardness of crystals and minerals. It runs from 1 to 10, with those measured as 1 being the softest, and those measured as 10 being the hardest.

# AMAZING CRYSTAL PLACES

Crystals can be found all over the world, but they are often well hidden. Here are some amazing places that are filled with crystals!

### Hang Son Doong cave, Vietnam

This 3-million-year-old **limestone** cave features some of the world's biggest **stalactites** and **stalagmites**. To keep it safe, only 1,000 visitors are allowed inside the cave each year.

### Surrounding Mt. Kilimanjaro, Tanzania

Tanzanite was discovered in the 1960s in Tanzania, Africa. The crystal is so **rare** that it has never been found anywhere else! It's thought to have formed from the heat and Earth movements that created Mount Kilimanjaro.

# Brandberg Mountain, Namibia

The areas around Brandberg mountain in the Namib desert, Africa, are known for their incredible amethyst crystals! Amethysts found here are called Brandberg Amethysts, and are a smoky purple colour.

# Minas Gerais, Brazil

The area of Minas Gerais in Brazil has some of the richest crystal deposits in the world. It's famous for producing lots of incredible, high-quality gems, including emeralds, aquamarine, and topaz.

# WHAT ARE GEMSTONES?

A gemstone is a mineral or crystal that has been cut and polished. Gemstones are usually very expensive.

## FROM CRYSTALS TO GEMSTONES

Crystals are turned into gemstones when they are cut and polished into perfect, sparkly shapes. Many gemstones are made from very rare crystals, which is why they are also called precious stones.

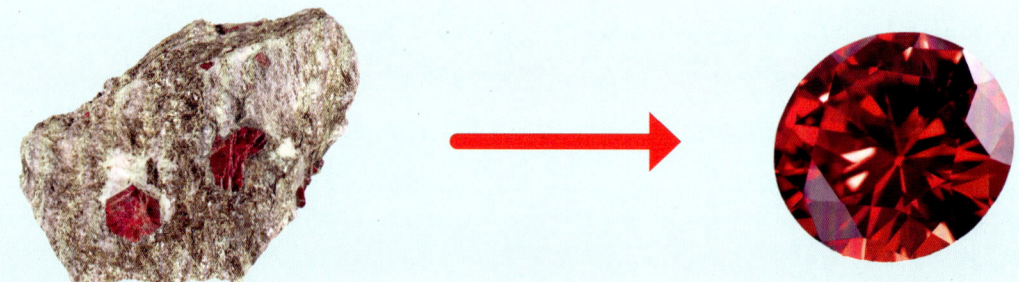

**Red corundum** can be cut and polished to make **RUBIES.**

**Beryl** can be cut and polished to make **EMERALDS.**

**Olivine** can be cut and polished to make **PERIDOT.**

# ORGANIC GEMSTONES

Some gemstones are formed from animals or plants. These are called **organic** gemstones. Unlike crystals, they can't be tested for hardness.

## AMBER

Amber is made of tree **resin**. Resin can seep out of the bark of trees and slowly harden. If it hardens and survives over millions of years, it turns into amber.

## CORAL

Coral is made by millions of tiny creatures that live in tropical oceans. Coral can be a variety of bright colours and interesting patterns, which makes it great for jewellery.

## PEARL

Sometimes grains of sand get trapped inside the shells of shellfish, like oysters. The animal coats the sand with tiny crystals. Over time, the sand turns into a pearl!

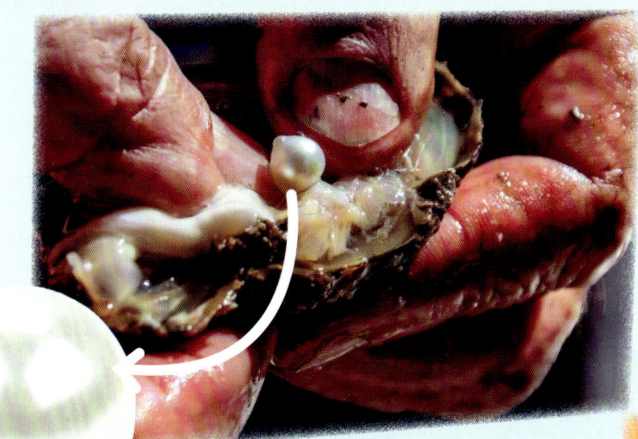

# HOW WE USE CRYSTALS AND GEMSTONES

For centuries, crystals and gemstones have been very important in many cultures. They can be a sign of power and wealth. Some people believe they can bring good luck, too.

## HEALING POWERS

Ancient Egyptians believed that some gemstones could heal people. There is no scientific evidence to prove that it works, but some people today still use crystals and gemstones to try and cure illnesses.

## ROYAL GEMS

Diamonds are the hardest and most precious gemstones. For centuries, they have been a symbol of wealth, and set in royal crowns and precious jewellery.

## MODERN TECHNOLOGY

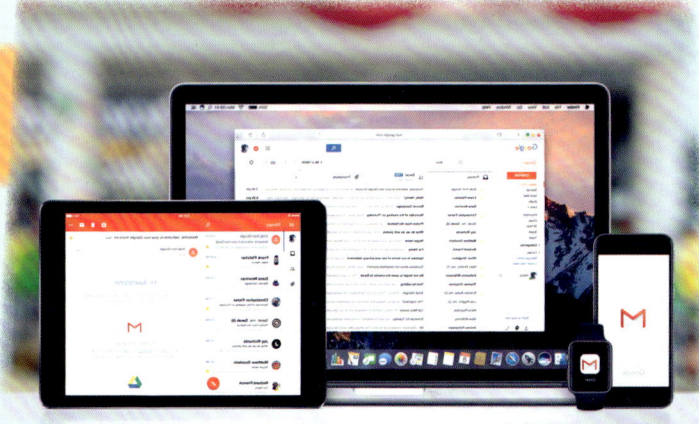

Some crystals, such as quartz, stay strong and stable under pressure, which makes them useful in building technology. Crystals help clocks and watches keep time, conduct electricity in devices like solar panels, and make screens on laptops and tablets last longer.

## DRILLING DOWN

Crystals may look delicate, but that's not always true. Being one of the hardest materials in the world makes diamonds perfect for drilling into super hard materials like rock! They break through much quicker and easier than drills made of metal.

17

# GEMSTONE DETECTIVES

**There's so much to learn about gemstones, but how do we know about them and the secrets they hold?**

The scientists who study gemstones are called gemologists. It's thanks to their studies that we have learnt so much about gems and the features that make each one unique.

Gemologists spend most of their time in the **LAB**. They do tests on gems, examining their cut, colour, quality, and clarity to identify different gems. For those that are harder to tell apart, like rubies and garnets, they look at their physical properties, for example by testing how they react to light.

Part of a gemologist's job can include working out whether gems are natural or not. Today, lab-grown gems can be so realistic that they can only be told apart with the help of machines like microscopes and spectrometers. Gemologists work with jewellery suppliers, gem cutters, and metalworkers to ensure they have the information they need to work with gems.

# COLLECTING CRYSTALS

These shiny materials are collected by people all over the world. Different people prefer different crystals. Here's some tips for starting your own crystal collection.

## BIRTHSTONES

Many countries have a tradition of birthstones. These are gemstones that each represent a month of the year. Many people collect and wear the gemstone of the month they were born in. They believe this will bring them luck.

| JANUARY GARNET | FEBRUARY AMETHYST | MARCH AQUAMARINE | APRIL DIAMOND | MAY EMERALD | JUNE PEARL |
| JULY RUBY | AUGUST PERIDOT | SEPTEMBER SAPPHIRE | OCTOBER OPAL | NOVEMBER TOPAZ | DECEMBER TURQOISE |

What's your birthstone?

## NATURAL CRYSTALS

Some people prefer to collect a range of crystals to display in their homes. These can be cut into special prism-like shapes, polished into smooth pebble-like shapes, or be in their natural form.

## SEARCHING FOR CRYSTALS

The best way to start a crystal collection is by visiting local rock shops, where you can see the crystals in person. Always go with an adult, and ask them to make sure that the crystal is real before buying it.

## DISPLAYING YOUR CRYSTAL COLLECTION

Crystals are delicate and cleaning them is difficult, so it's best to keep them away from dust. If they do need cleaning, use a soft paint brush and be careful! Why not make a display box for them? Use a glass-fronted box or cabinet, and use card to write the name of each type of crystal and where it was found.

# COMMON CRYSTALS

As we have seen, there are lots of different crystals. Here are some of the most common ones.

### Feldspar
- Colour: White, pink, or grey
- Hardness: 6
- Found in: Igneous and metamorphic rock

### Tourmaline
- Colour: Any, but black is the most common
- Hardness: 7
- Found in: Igneous rock

### Mica
- Colour: Black, white, green, or red
- Hardness: 2½
- Found in: Igneous, metamorphic, and sedimentary rock

### GYPSUM

- Colour: Mainly white, but can be grey, brown, or pink
- Hardness: 2
- Found in: Sedimentary rock

### MILKY QUARTZ

- Colour: White
- Hardness: 7
- Found in: Igneous rock and **mineral veins**

### HALITE (ROCK SALT)

- Colour: White, pink, orange, or blue
- Hardness: 2½
- Found in: Sedimentary rock

# ICONIC GEMSTONES

People have found many different and beautiful gemstones across the world. Here are some of the most famous gemstones of all.

### EMERALD

- Colour: Green
- Hardness: 8
- Found in: Igneous, metamorphic, and sedimentary rock

### DIAMOND

- Colour: Most commonly clear
- Hardness: 10
- Found in: Igneous rock

### BLUE SAPPHIRE

- Colour: Blue
- Hardness: 9
- Found in: Igneous and metamorphic rock

## RUBY

- Colour: Red
- Hardness: 9
- Found in: Igneous and metamorphic rock

## AQUAMARINE

- Colour: Pale blue or green-ish blue
- Hardness: 8
- Found in: Igneous and metamorphic rock

## JADE

- Colour: Mainly green, but can be purple, red, yellow, white, or black
- Hardness: 7
- Found in: Metamorphic rock

# UNIQUE CRYSTALS & GEMSTONES

Crystals and gemstones can be weird and wonderful too!

## Larimar

- Notable feature: A rare variety of pectolite that looks like the ocean
- Hardness: 5
- Found in: Igneous rock

## Autunite

- Notable feature: Glows under **ultraviolet light**
- Hardness: 2
- Found in: Igneous rock and mineral veins

## Mimetite

- Notable feature: Forms in "grape-like" bubble clusters.
- Hardness: 4
- Found in: Igneous, metamorphic, and sedimentary rock

# Agate

- Notable feature: A type of quartz that forms unique patterns and colours
- Hardness: 7
- Found in: Igneous, metamorphic, and sedimentary rock

# Dumortierite quartz

- Notable feature: Blue dumortierite needle-like crystals encased in clear quartz
- Hardness: 7
- Found in: Metamorphic rock

# Rainbow fluorite

- Notable feature: Mixture of different colours within one chunk of fluorite crystal
- Hardness: 4
- Found in: Igneous, metamorphic, and sedimentary rock

# RECORD BREAKERS

So many amazing crystals have been found before, but these ones are some of the most impressive.

---

The **BIGGEST** crystals ever found on Earth are in Mexico, in a place called the "Cave of the Crystals". They are made of gypsum and are around 11 metres (36 ft) long. That's the same length as a school bus!

You may think of diamonds as colourless, but they can be different colours. One of the **MOST PERFECT** diamonds ever was a pink diamond that was discovered in 1999. It was **flawless** inside and sold for millions!

28

There are many **RARE CRYSTALS** in the world, but one of the most impressive is red beryl. It's only found in the USA and was originally called "bixbite". Only one red beryl crystal is found for every 150,000 diamonds!

The **AMBER ROOM** was built in 1711 for the king of Prussia. It was decorated with amber panels and gold leaf, and took 10 years to build! The Amber Room was destroyed in World War II, but has since been rebuilt. It's in the Catherine Palace in St Petersburg, Russia.

The crystal Alexandrite has a **UNIQUE QUALITY** – it changes colour in different lights. In daylight, it looks blue or green, and in lower light, it looks red or purple! It is an extremely rare crystal.

# TRUE OR FALSE?

There is lots to learn about crystals and gemstones. How well do you know your facts?

### Crystals can completely fill rocks!

**TRUE!** Some crystals form into geodes – they completely fill holes in rocks and make little crystal caves! Quartz, like amethyst, are the most common geode crystals.

### Quartz is one of the rarest crystals on Earth!

**FALSE!** There are many different types of quartz, which range in colour from clear to black!

**Star rubies and star sapphires are very common!**

**FALSE!** Star rubies and sapphires are incredibly rare, which makes them very valuable. These gems reflect the shape of a star when light is shone on them.

**Some gems are called "cat's eyes"!**

**TRUE!** They are cut in a special way so that when light is shone on them, it is reflected in a single bright line, making them look just like a cat's eye!

**Some of the world's most precious crystals are kept in museums!**

**TRUE!** The Hope Diamond is one of the biggest diamonds in the world. It's kept in the Smithsonian Institute museum in the USA.

# GLOSSARY

**Evaporates** – when a liquid turns into a gas.

**Flawless** – something that is perfect.

**Igneous rocks** – a type of rock that forms when molten rock cools down and hardens.

**Limestone** – a type of sedimentary rock that is made mostly of animal remains, shells, and mud.

**Metamorphic rocks** – a type of rock that forms when heat or pressure causes rocks to change their structure or mineral composition.

**Minerals** – substances that are naturally found in things like rocks, sand, and soil. Many minerals form as crystals.

**Mineral veins** – distinct layers or stretches of minerals that run through rocks.

**Mohs Hardness Scale** – a scale which is used to measure the hardness of minerals.

**Molten** – another word for melted.

**Organic** – something that is made from a natural source like a plant or animal.

**Pressure** – a force that presses on something.

**Rare** – something that is uncommon

**Resin** – a thick, sticky substance that comes from trees.

**Sedimentary rocks** – a type of rock that forms when small, worn off pieces of other rocks become joined together in layers.

**Stalactites** – formations that are made from minerals, but look like icicles and grow down from cave ceilings.

**Stalagmites** – formations that are made from minerals, but look like icicles that are growing up from cave floors. They are usually underneath stalactites (see above).

**Ultraviolet light** – a strong and powerful type of light.

**Unique** – something that is one of its kind; unlike all others.

# INDEX

**A**
Agate 27
Alexandrite 29
Amber 15, 29
Amethyst 13, 17, 30
Apatite 11
Aquamarine 13, 17, 25
Autunite 26

**B**
Beryl 14, 29
Brandberg Mountain, Namibia, Africa 13

**C**
Calcite 11
Chrysoprase 14
Coral 15
Corundum 11, 14

**D**
Diamond 11, 17, 24, 28-29, 31

**E**
Emerald 13, 14, 17, 24

**F**
Feldspar 11, 22
Fluorite 8, 11, 27

**G**
Garnet 17, 18
Gemologists 18-19
Gemstones 5, 10, 14-15, 16-17, 18-19, 24-25, 26-27
Gneiss 6
Gold 21, 29
Granite 6, 9
Gypsum 9, 11, 23, 28

**H**
Halite 8, 23
Hang Son Doong cave, Vietnam 12
Haematite 9

**I**
Ice 5, 7
Igneous rock 6-7, 22-23, 24-25, 26-27

**J**
Jade 25

**L**
Larimar 26
Limestone 12

**M**
Malachite 10
Metamorphic rock 6, 22, 24-25, 26-27
Mica 9, 22
Mimetite 26
Minas Gerais, Brazil 13
Minerals 6, 11, 14, 23, 26
Mount Kilimanjaro, Tanzania, Africa 12

**O**
Olivine 14
Opal 17

**P**
Pearl 15, 17
Peridot 14, 17
Pyrite 4

**Q**
Quartz 4, 10-11, 16, 23, 27, 30

**R**
Ruby 14, 17, 18. 25, 31

**S**
Salt 5, 7, 23
Sapphire 10, 17, 24, 31
Sedimentary rock 6-7, 22-23, 24, 26-27
Stalactites 12
Stalagmites 12

**T**
Talc 11
Tanzanite 12
Topaz 10-11, 13, 17
Tourmaline 22
Turquoise 17

**U**
Uses of crystals and gemstones 5, 10, 16